JOURNEY TO THE FUTURE OF TRANSPORT

BY AILYNN COLLINS
ILLUSTRATED BY DANIEL PEDROSA

raintree
a Capstone company — publishers for children

Raintree is an imprint of Capstone Global Library Limited, a company incorporated in England and Wales having its registered office at 264 Banbury Road, Oxford, OX2 7DY – Registered company number: 6695582

www.raintree.co.uk
myorders@raintree.co.uk

Copyright © Capstone Global Library Limited 2024
The moral rights of the proprietor have been asserted.

All rights reserved. No part of this publication may be reproduced in any form or by any means (including photocopying or storing it in any medium by electronic means and whether or not transiently or incidentally to some other use of this publication) without the written permission of the copyright owner, except in accordance with the provisions of the Copyright, Designs and Patents Act 1988 or under the terms of a licence issued by the Copyright Licensing Agency, 5th Floor, Shackleton House, 4 Battle Bridge Lane, London, SE1 2HX (www.cla.co.uk). Applications for the copyright owner's written permission should be addressed to the publisher.

British Library Cataloguing in Publication Data
A full catalogue record for this book is available from the British Library.

ISBN 978 1 3982 5477 0

Editorial Credits
Editor: Aaron Sautter
Designer: Elyse White
Media Researcher: Rebekah Hubstenberger
Production Specialist: Whitney Schaefer

Originated by Capstone Global Library Ltd

All the internet addresses (URLs) given in this book were valid at the time of going to press. However, due to the dynamic nature of the internet, some addresses may have changed, or sites may have changed or ceased to exist since publication. While the author and publisher regret any inconvenience this may cause readers, no responsibility for any such changes can be accepted by either the author or the publisher.

Printed and bound in India

CONTENTS

SECTION 1:
ARE ELECTRIC CARS THE FUTURE? 6

SECTION 2:
OFF TO THE FUTURE 10

SECTION 3:
LONG DISTANCE GROUND TRANSPORT 18

SECTION 4:
THE FUTURE OF FLIGHT 22

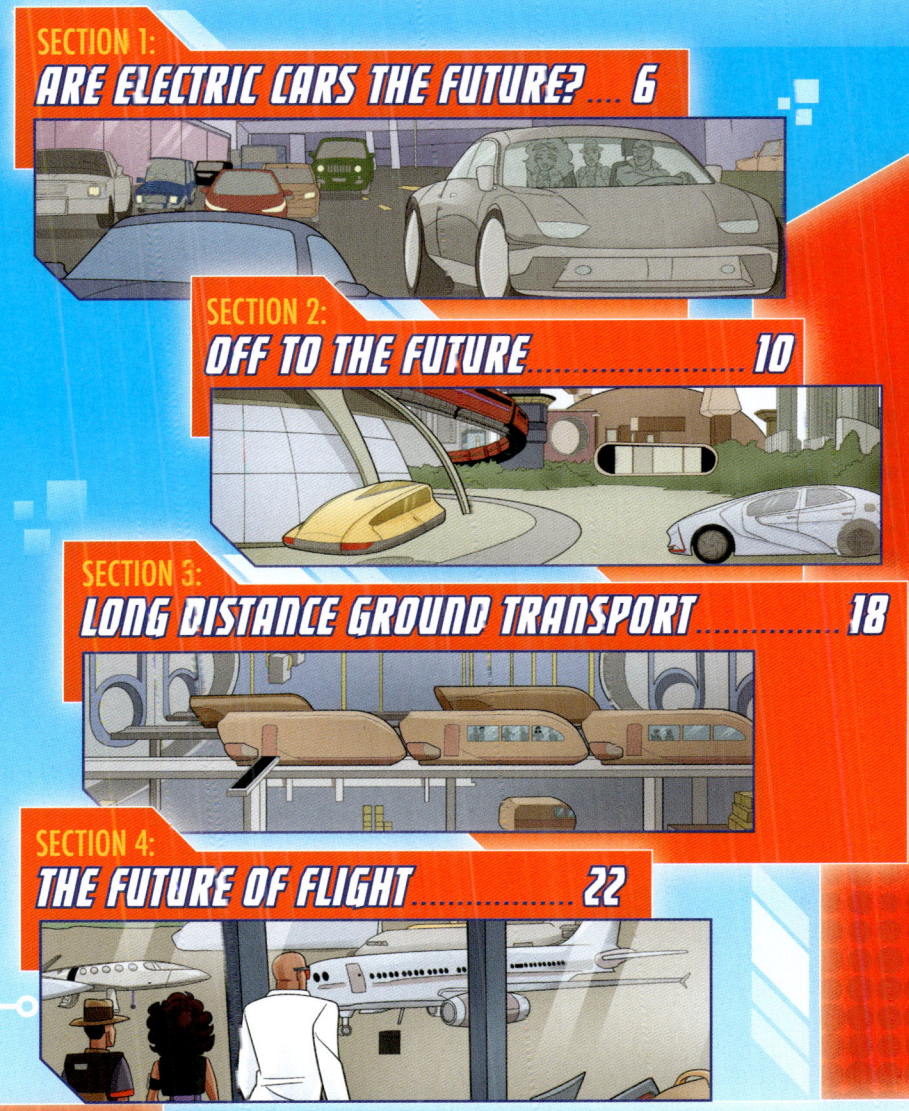

MORE ABOUT THE FUTURE OF TRANSPORT	28
GLOSSARY	30
FIND OUT MORE	31
ABOUT THE AUTHOR	32
ABOUT THE ILLUSTRATOR	32

THE SOCIETY OF SUPER SCIENTISTS

MAX AXIOM

After years of study, Max Axiom, the world's first Super Scientist, knew the mysteries of the universe were too vast for one person alone to uncover. So Max created the Society of Super Scientists! Using their superpowers and super-brains, this talented group investigates today's most urgent scientific and environmental issues and learns about actions everyone can take to solve them.

LIZZY AXIOM

NICK AXIOM

SPARK

THE DISCOVERY LAB

Home of the Society of Super Scientists, this state-of-the-art lab houses advanced tools for cutting-edge research and radical scientific innovation. More importantly, it is a space for Super Scientists to collaborate and share knowledge as they work together to tackle any challenge.

SECTION 1: ARE ELECTRIC CARS THE FUTURE?

IS NO TRAFFIC POSSIBLE?

One way to relieve traffic jams is by encouraging people to use public transport. Buses, underground trains and light rail systems could carry more people than cars can. This would help reduce the number of cars on roads as well as the pollution they create.

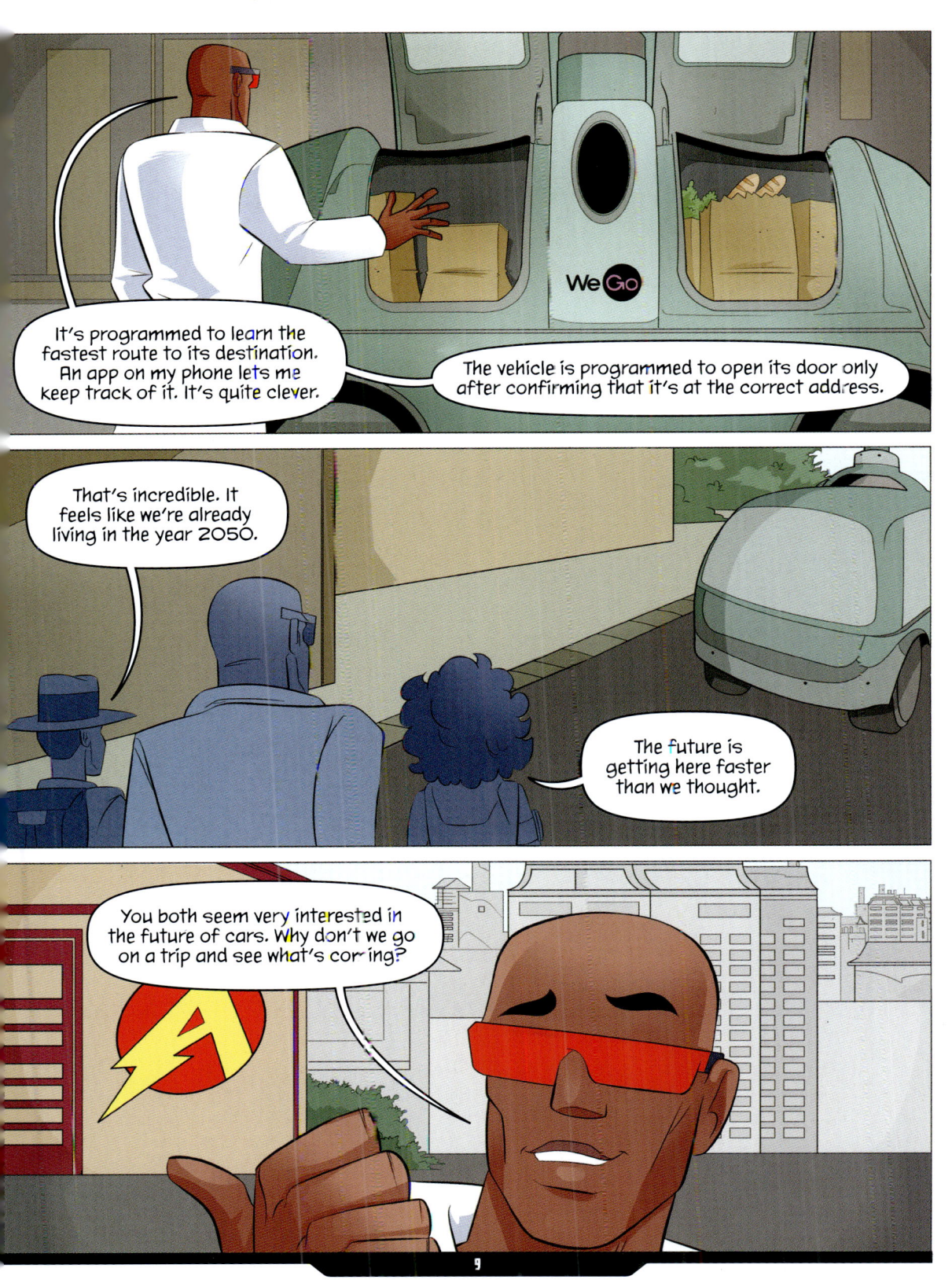

SECTION 2: OFF TO THE FUTURE

"We're going to the future?"

"I've made some changes to the portal. We can visit the year 2050 now."

"Let's see what the future of transport really looks like."

"Welcome to 2050!"

"It's a lot quieter here than in our time. Is it because the vehicles are electric?"

ZOOOOP!

"Some of them are. Some are solar-powered. But all use renewable energy."

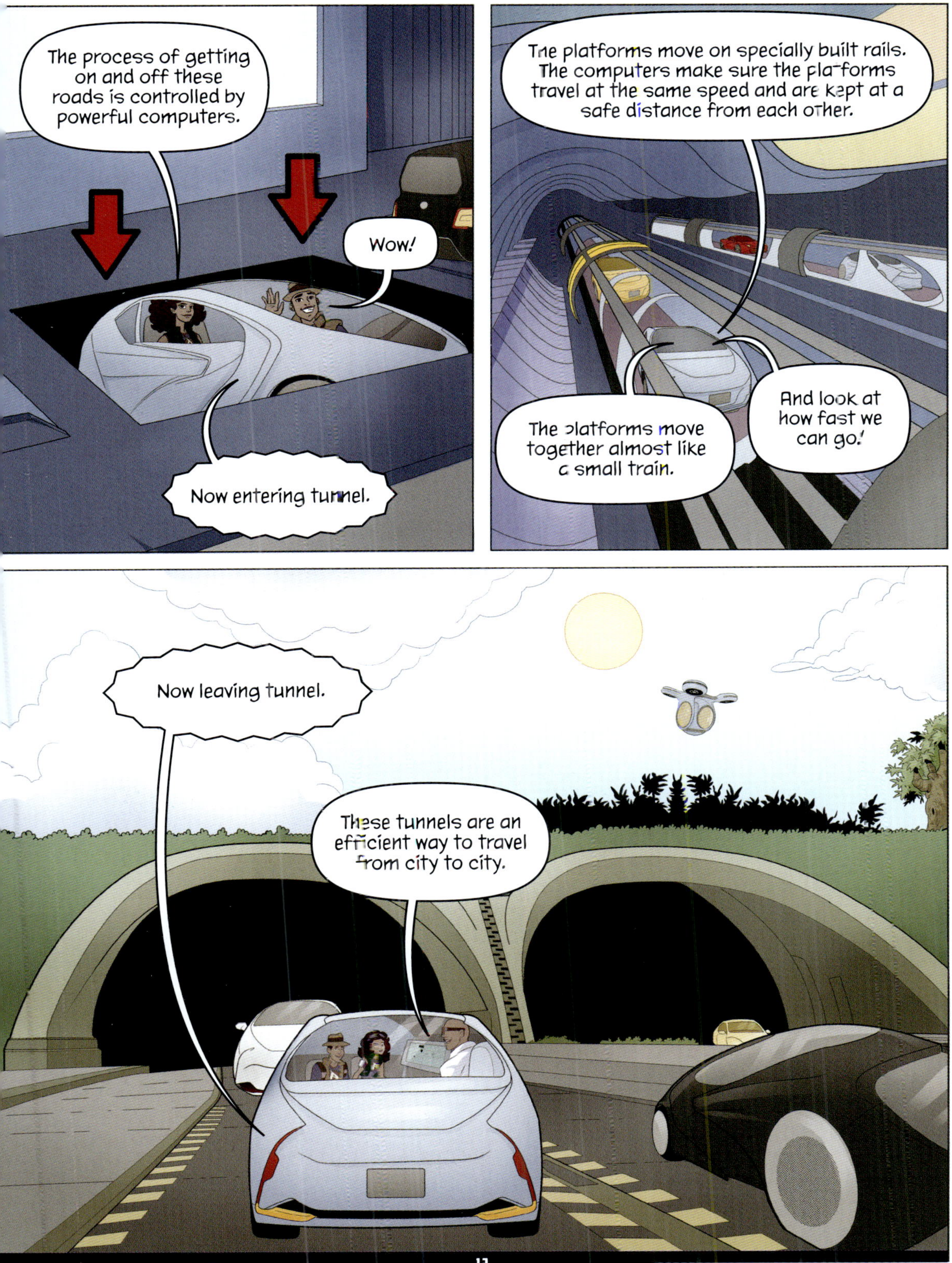

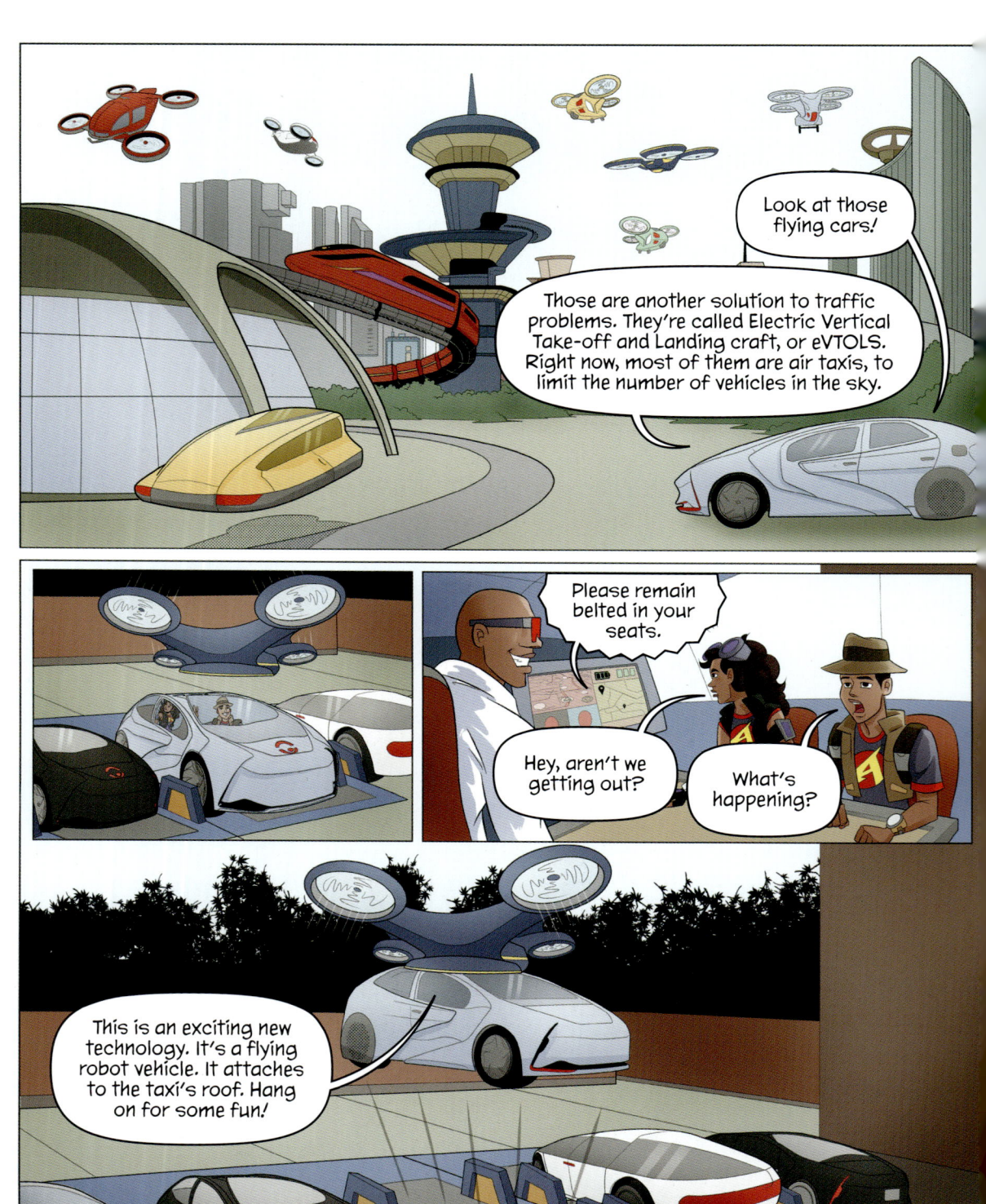

WHAT IF THERE'S A POWER CUT?

One of the questions transport scientists and engineers must answer is what to do in a power cut. Today's electric vehicles go into "turtle mode". They slow down to 30 kilometres per hour and pull over to the side of the road. In the future, improved batteries will last much longer. Vehicles may also have solar panels to create enough power to get to safety.

SECTION 3: LONG DISTANCE GROUND TRANSPORT

This is a Maglev train. That's short for Magnetic Levitation.

This train doesn't have an engine. Instead, a magnetic field lifts the train off the track and pushes it forwards.

Because there's no friction, the train moves very smoothly and fast.

I remember doing an experiment with magnets in science class. Magnets attract and repel each other. We floated a magnetic toy car above a magnet with the same poles lined up with each other.

RESISTING THE AIR

If you stick your hand out the window of a moving car, you can feel the wind pushing you back. This is air resistance, or friction. In a vacuum, there is no air to push back on an object. A vehicle inside a vacuum can travel very fast with no resistance.

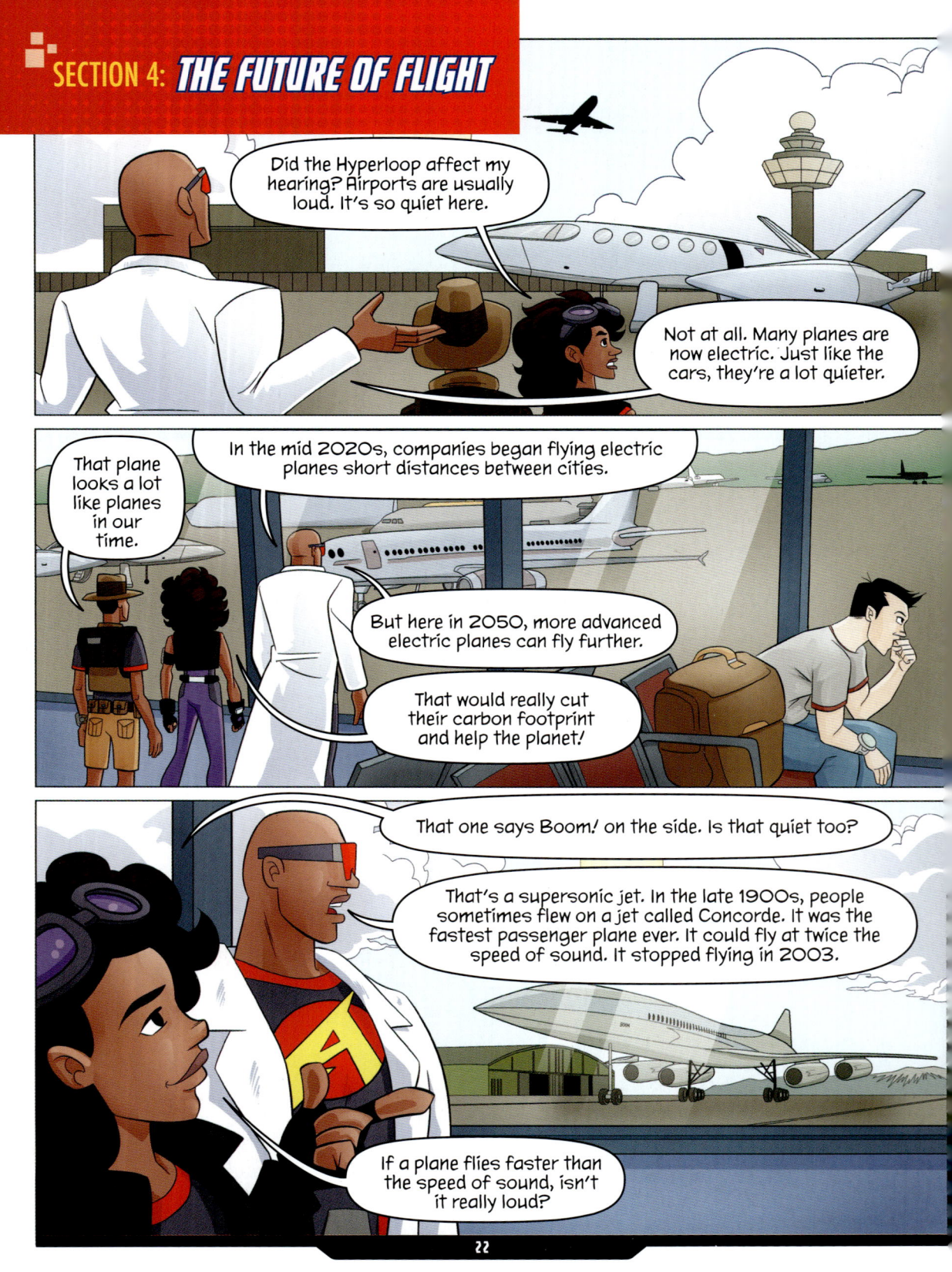

Yes, when an object, like an aeroplane, moves faster than the speed of sound, it creates a shock wave. It sounds like loud thunder or an explosion. This is called a sonic boom.

But these advanced supersonic planes have new designs that reduce the effects of sonic booms.

They also don't go supersonic until they're high in the air or over water, where there aren't any homes. The boom can be spread out in a wider area, making it less noisy.

This says that in 2030, a company called Hermeus launched a hypersonic passenger plane. It flew from New York to Paris in just 90 minutes.

What's that one?

It flew as fast as five times the speed of sound. Passengers felt themselves pushed back into their seats for about ten minutes. When the plane reached above 30,000 metres, those effects wore off, since the air is so thin up there.

That's a hydrogen-powered plane. It uses hydrogen just like we use petrol in cars. But it has to carry a lot more fuel, so the wing design had to change.

A HYDROGEN-FUELLED PLANE?

Hydrogen can be cooled and stored in liquid form. Then it can be used to generate electricity that would power a plane. It could also be used like jet fuel to power the plane directly. There are still several dangers to overcome, but scientists are working to make the hydrogen plane a reality. If successful, it could help reduce the need for using fossil fuels.

"What's that over there?"

"That's something new, even in 2050. It's a rocket used for sub-orbital flights."

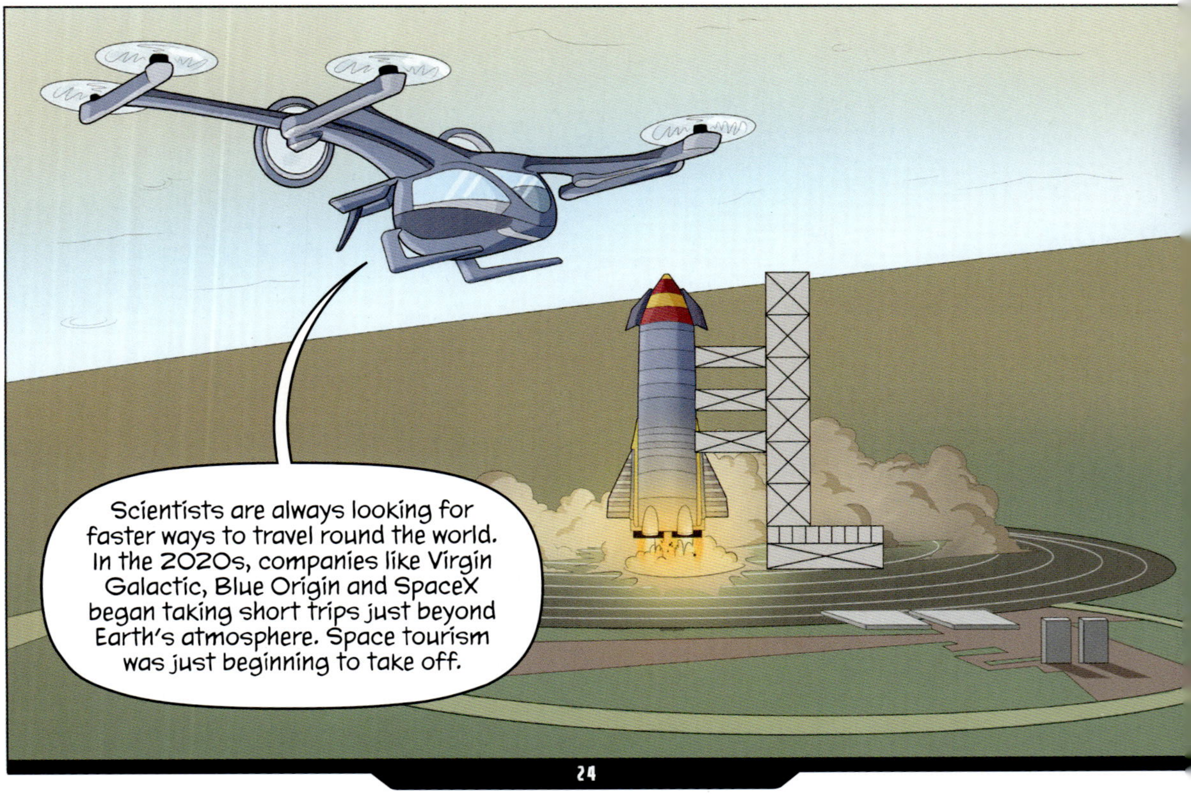

Scientists are always looking for faster ways to travel round the world. In the 2020s, companies like Virgin Galactic, Blue Origin and SpaceX began taking short trips just beyond Earth's atmosphere. Space tourism was just beginning to take off.

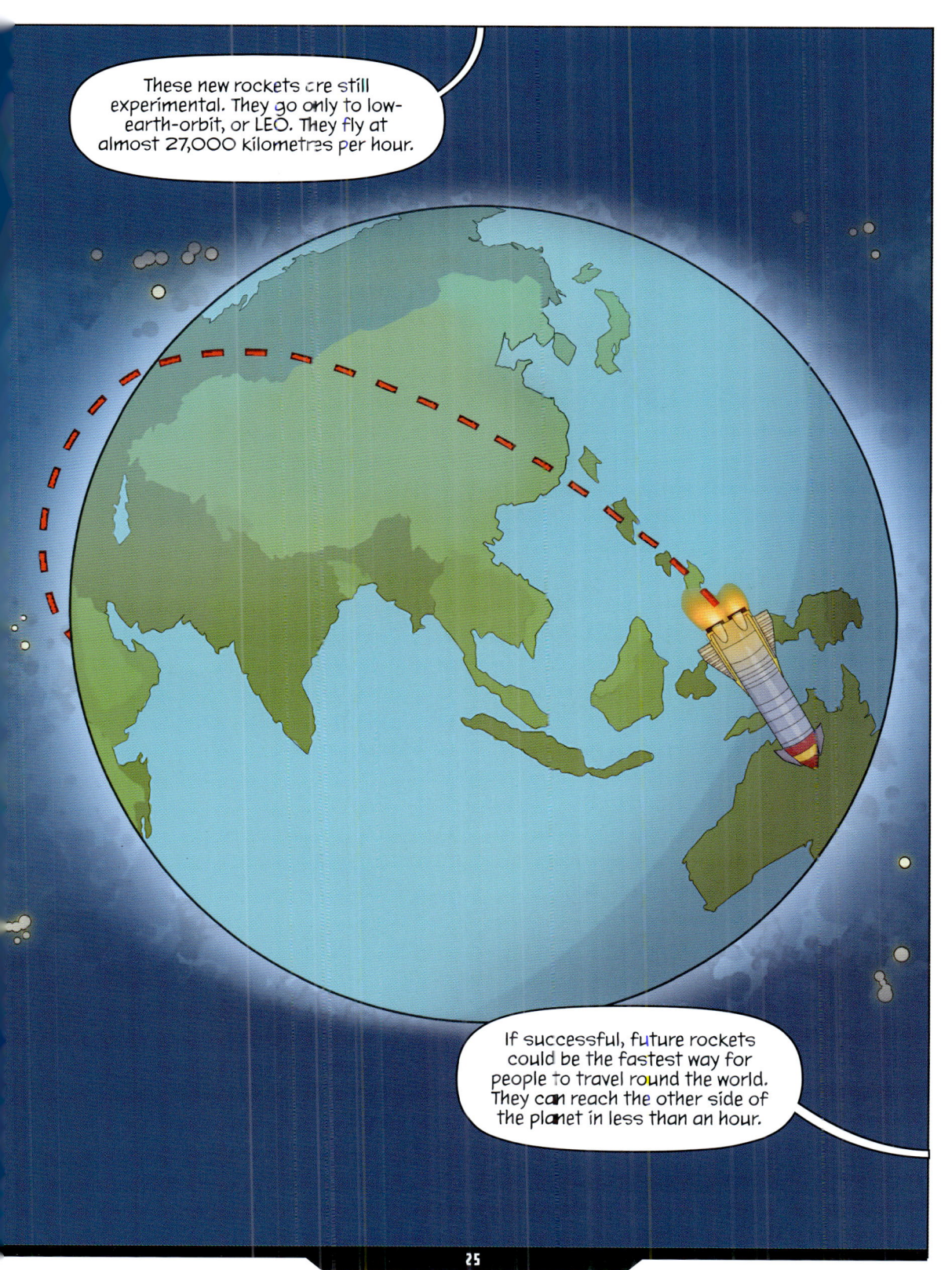

"I'm guessing those ships and boats in the water run on electricity too, right?"

"And they're also controlled by computers."

"You're both right. The new technology used for cars and planes is used for ships too."

"Come on, it's time to get home."

ZOOOOP!

"I'm going to miss those flying cars."

"I can't wait to see the future become a reality."

MORE ABOUT THE FUTURE OF TRANSPORT

Many new cars today have computers on board to help humans drive more safely. Some can sense objects around them and slow down, or even stop when they get too close to another vehicle.

Some vehicles can suggest that the driver takes a break during a long trip. Others also have a version of autopilot, where the driver only needs to keep a hand on the steering wheel. The car travels at the correct speed, at a safe distance from the car in front.

Today, many vehicles have computer systems that can provide maps and real-time traffic news. They are powered by GPS and can direct the driver to take the quickest route to their destination.

Completely autonomous cars, where there is no human driver at all, are still experimental. Some have been in accidents that forced companies to remove the self-driving cars from the road.

Countries like the UK, United States, Japan, Germany and South Korea are closer to having truly self-driving cars. They have better roads and greater support from their governments. But in the near future, self-driving cars may be a common sight on most roads.

Several companies have already begun using robotic delivery vehicles. Some travel along on pavements, while others, such as delivery drones, fly overhead. Companies like Alibaba in China and Amazon.com are delivering packages to residents in cities without a human driver. These automated vehicles could soon become a regular sight. In some ways, the future of transport has already arrived.

GLOSSARY

artificial intelligence ability of a computer or machine to think and behave like a human

automated use of machines to do work instead of people

autonomous something, such as a vehicle or machine, that can control itself rather than be controlled by a person

autopilot system on a vehicle that allows it to steer and control itself

carbon footprint amount of carbon dioxide released into the atmosphere by a person's activities

drag force that resists the motion of an object moving through air

friction force created when two objects rub together; friction slows down objects

GPS short for Global Positioning System; an electronic tool used to find the location of an object or directions to a certain location

gyroscopic ability to turn freely in any direction while maintaining balance

hydroelectric making electric power from the force of moving water

radar device that uses radio waves to track the location of objects

renewable energy power from sources that will not be used up, such as wind, water and the Sun

FIND OUT MORE

BOOKS

Artificial Intelligence at Home and on the Go (The World of Artificial Intelligence), Tammy Enz (Raintree, 2020)

The Tech Behind Electric Cars (Tech on Wheels), Matt Chandler (Raintree, 2020)

What Would it Take to Build a Flying Car? (Sci-Fi Tech), Megan Ray Durkin (Raintree, 2020)

WEBSITES

www.bbc.co.uk/newsround/62604275
Read about the government's plans to introduce self-driving cars on the roads by 2025.

www.bbc.co.uk/newsround/65602284
Learn more about the future of international travel through space flight.

www.dkfindout.com/uk/transport/history-cars/flying-car/
Learn more about flying cars with DKFindout!

ABOUT THE AUTHOR

Ailynn Collins has written several non-fiction children's books about amazing people, space and science. Ailynn also loves to write fiction, especially stories about aliens, ghosts, witches, dinosaurs and travelling through the universe. She lives outside Seattle, Washington, USA, with her husband and five dogs.

ABOUT THE ILLUSTRATOR

Daniel Pedrosa was born in Araraquara, a quiet town in São Paulo, Brazil. He has always had an interest in art. At the age of ten, his older brother gave him his first superhero comic. It was then that he decided that drawing comics would be his profession. In 2010, he began his professional career doing comic strips for newspapers and drawing children's supplements for tabloids and magazines. Today, Daniel creates colourful art for Criptozoik, Tildawave, Raintree, 137 Studios and produces advertising material for the largest Honda Motors store in Brazil.